TOOLS FOR TEACHERS

- **ATOS:** 0.6
- **GRL:** B
- **WORD COUNT:** 55

- **CURRICULUM CONNECTIONS:** animals, habitats

Skills to Teach

- **HIGH-FREQUENCY WORDS:** do, in, the, we, who
- **CONTENT WORDS:** bears, beetles, deer, forest, foxes, mice, owls
- **PUNCTUATION:** question marks, periods, exclamation point
- **WORD STUDY:** r-controlled vowels (*bears*, *forest*); plural endings s (*bears*, *beetles*, *owls*), es (*foxes*)
- **TEXT TYPE:** information report

Before Reading Activities

- Read the title and give a simple statement of the main idea.
- Have students "walk" though the book and talk about what they see in the pictures.
- Introduce new vocabulary by having students predict the first letter and locate the word in the text.
- Discuss any unfamiliar concepts that are in the text.

After Reading Activities

The book's text tells us that each animal lives in a forest, but the photos provide extra information. Encourage children to talk about the different things animals are shown doing in the book. Write the book's language pattern on the board: "Who (lives) in the forest?" in one column, and "____ do." in another. Based on the photos and prior knowledge, what other words could be used in place of "lives?" For example, "Who climbs in the forest?" and "Bears do.," or "Who flies in the forest?" and "Owls do." Write the children's answers under the appropriate column.

Tadpole Books are published by Jump!, 5357 Penn Avenue South, Minneapolis, MN 55419, www.jumplibrary.com

Editorial: Hundred Acre Words, LLC **Designer:** Anna Peterson

Photo Credits: Getty: Leo Leo, 4–5; Matthias Breiter, 2–3. Shutterstock: DutchScenery, 1; Eric Isselee, cover, 6–7; Gemenacom, 14–15; irin-k, 12–13; michaeljung, 14–15; Neil Burton, 10–11; ppa, 14–15; Przemyslaw Wasilewski, 10–11. SuperStock: Raimund Linke/Exactostock-1598, 8–9.

Library of Congress Cataloging-in-Publication Data
Names: Fretland VanVoorst, Jenny, 1972– author.
Title: Who lives in the forest? / by Jenny Fretland VanVoorst.
Description: Minneapolis, Minnesota: Jump!, (2017) | Series: Who lives here? | Audience: Ages 3–6. | Includes index.
Identifiers: LCCN 2017023540 (print) | LCCN 2017022507 (ebook) | ISBN 9781624967269 (ebook) | ISBN 9781620319550 (hardcover: alk. paper) | ISBN 9781620319567 (pbk.)
Subjects: LCSH: Forest animals—Juvenile literature.
Classification: LCC QL112 (print) | LCC QL112 .F36 2017 (ebook) | DDC 591.73—dc23
LC record available at https://lccn.loc.gov/2017023540

WHO LIVES IN THE FOREST?

by Jenny Fretland VanVoorst

TABLE OF CONTENTS

tadpole
books

WHO LIVES IN THE FOREST?

Who lives in the forest?

Bears do.

Who lives in the forest?

owl

Owls do.

Who lives in the forest?

Mice do.

Who lives in the forest?

deer

Deer do.

Who lives in the forest?

Foxes do.

Who lives in the forest?

Beetles do.

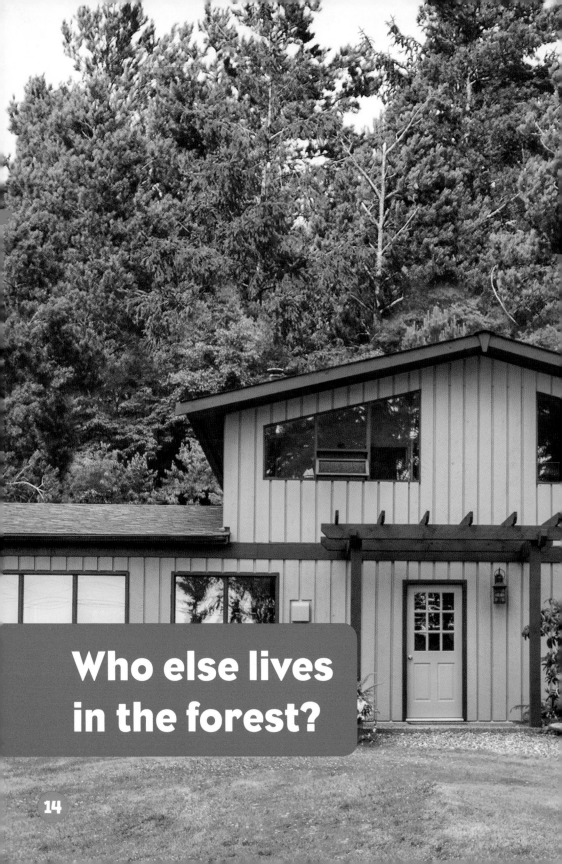

Who else lives in the forest?

We do!

15

WORDS TO KNOW

bears

beetles

deer

foxes

mice

owls

INDEX